Builders Iron Foundry

The Venturi Meter Patented by Clemens Herschel, Hydraulic Engineer and Builders Iron Foundry

Made by Builders Iron Foundry, Founders and Machinists

Builders Iron Foundry

The Venturi Meter Patented by Clemens Herschel, Hydraulic Engineer and Builders Iron Foundry
Made by Builders Iron Foundry, Founders and Machinists

ISBN/EAN: 9783337060039

Printed in Europe, USA, Canada, Australia, Japan

Cover: Foto ©berggeist007 / pixelio.de

More available books at **www.hansebooks.com**

THE VENTURI METER

PATENTED BY

CLEMENS HERSCHEL

Hydraulic Engineer

AND

BUILDERS IRON FOUNDRY

MADE BY

BUILDERS IRON FOUNDRY

FOUNDERS AND MACHINISTS

.

Providence, R. I., u. s. a.
1898.

PREFACE

In the papers that have hitherto been published by us concerning the Venturi Meter, (copies of which will be furnished upon application), we have given a more or less technical explanation of the physical laws governing the action of the meter, and called attention to the uses to which it could be applied. The object of this pamphlet is to again state the facts relating to the operation of the meter, as briefly as possible, and to show by some illustrations how it has been applied in actual uses.

BUILDERS IRON FOUNDRY.

PROVIDENCE, R. I., January 1, 1898.

THE VENTURI METER

The Meter is named for the Italian philosopher Origin of name. Venturi, who first called attention, in 1796, to the relation between the velocities and pressures of fluids when flowing through converging and diverging tubes.

The Meter consists of two parts—the Tube, Principal parts of meter. through which the water flows, and the Register, which sums up and indicates on a dial the quantity of water that has passed through the tube.

The Tube is formed of two truncated cones, The Tube, its shape. joined at their smallest diameters by a short throat piece. At the upstream end and at the throat there are encircling pressure chambers that are connected with the interior by carefully drilled holes, and from which pressure pipes lead to the register. See Figure 1.

The operation of the Venturi Meter is due to the Operation of the meter. fact that when water in any pipe passes from a state of rest to movement, or from one velocity of flow to a greater velocity, a certain amount of pressure against the shell of the pipe disappears, and that the disap- Velocities. pearance of pressure, or loss of head, is entirely dependent upon the velocities of flow past the points in the pipe at which pressure is taken.

Therefore, at two points in a taper pipe, or Venturi tube, as at U–T, Figure 1, because of different sec-

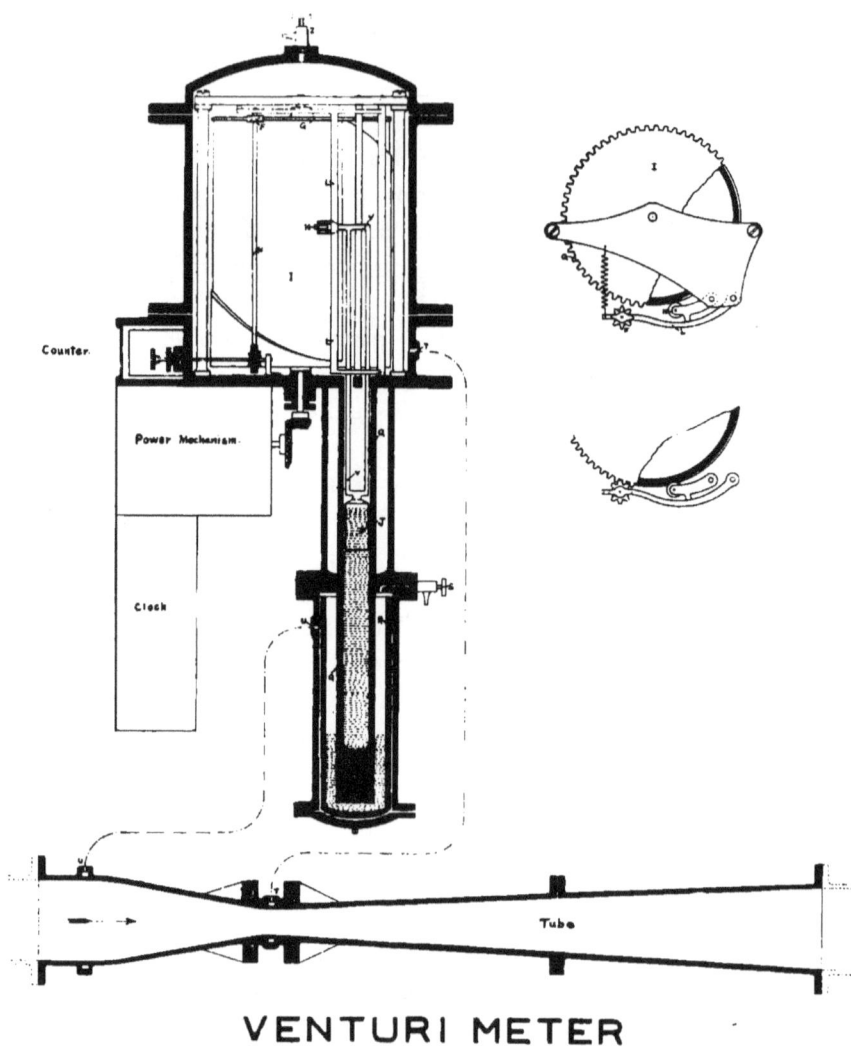

Counter.

Power Mechanism.

Clock

Tube

VENTURI METER

FIGURE 1.

SECTIONAL VIEW OF VENTURI METER TUBE AND REGISTER.

tional area, different velocities and consequently different pressures must exist whenever there is any flow
through the tube. The difference in pressure at the
two points is always the same for the same velocity
of flow, whatever the total or hydraulic pressure may
be; and by exhaustive experiment has been shown to
be nearly equal (in feet of water) to 1-64 the square
of the velocity of flow (in feet per second) through
throat of meter tube; or, in other words, to coincide
closely with the fundamental hydraulic formula for
the head corresponding to any velocity of discharge
from an orifice,

$$h = \frac{V^2}{2g}$$

in which "h" corresponds to the difference in pressure
at U and T, V the velocity of flow through throat,
and g the acceleration of gravity.

For demonstration of the preceding statements, see Herschel's
Rowland Prize Paper, Transactions of American Society of Civil
Engineers, December, 1877. Reprint furnished on application.
Merriman's Hydraulics, Article 71, (Reprinted herewith.)
Illustrations of the Theorem of Bernouilli under " Hydro-
mechanics,"—9th Edition Encyclopædia Britannica, or reprint
furnished on application, and almost any modern text book on
Hydraulics.

The different pressures existing at the upstream Register.
end and throat of the meter tube are transmitted by
small pipes T–U, to the register (Figure 1), where
they oppose one another, and are balanced by dis-
placement of level of two columns of mercury in
cylindrical tubes, one within the other. The inner
mercury column carries a float, J, V, the position
of which is dependent on, and as previously explained
is an indication of the velocity of water flowing through

7

FIGURE 2. REGISTER.

the tube. The position assumed by an idler wheel H carried by this float, relative to an intermittently revolving integrating drum I, determines the duration of contact of gears G and F connecting drum and counter, by which the flow for successive intervals is registered.

It is a common but erroneous impression that water flowing through a contracting pipe brings an increased pressure against the entire converging surface which it meets. The reverse of this impression is true. The pressure of water flowing through the Venturi Tube decreases from the inlet to the throat, and increases from the throat to the outlet. The difference between pressures at inlet and outlet ends of the Tube is the friction head or loss of head caused by its operation, and under ordinary circumstances is inconsiderable. The amount of this loss in tubes with throat area 1-9 of main is stated in the accompanying tables and shown by diagram, Figure 10. By adaptation of the tube to requirements, the loss of head may be limited to any desired amount.

There is no limit to the size of the meter tubes, nor the quantity of water that may be measured. The largest that has yet been made is 9 feet diameter, with maximum capacity at the rate of more than 200,000,000 gallons in 24 hours.

Usually the meter tubes, for sizes under 60 inches diameter, are made of cast iron, with bronze-lined throat pieces, but for special service may be made of wooden staves, sheet steel, cement-concrete, brick or other material, with suitable metal parts for throat and upsteam pressure chambers.

The tube is usually laid as a part of the pipe line and is not injuriously affected by water hammer or

9

FIGURE 3.

BACK OF REGISTER.

the most violent fluctuations of velocity or pressure, and requires no more care than the pipe line itself. The meter cannot be disarranged by fish, gravel or other substances carried through the pipe line by the water.

The meter may be said to have created a field of usefulness for water meters which did not previously exist. It accomplishes with little difficulty what otherwise is done only laboriously or approximately and clumsily.

General usefulness.

In water works, this meter enables a record to be kept of the total quantity consumed, also, of the quantities consumed by large users, such as adjacent towns and cities, the several districts of one and the same city, railroads, factories and the like. See Fig. 11.

Special advantages for water works.

As it cannot be disarranged by substances in the water, it is especially desirable, when the water it measures is liable to be used for fire service.

Fire service.

It can be used as a " waste-water meter," keeping a record of the quantity passing the meter at any time. Its use in the detection of wastes and leaks,* and as a measure of the slip of pumps,† and the action of filter plants, makes it very valuable to all works for a public supply of water.

A similar line of service can be done by this meter in the case of sewerage systems, many of which, as now built, are constructed and operated for the joint benefit of several towns and cities, with the cost of operation divided pro rata between them, according to the quantity of sewage contributed.

Special advantages for sewerage system.

For irrigation works this meter can accomplish what has hitherto been desired but has not been practicable. It enables water for irrigation purposes to be sold strictly by measure, and with practically no constraint as to the time when it may be drawn.

Special advantages for irrigation works.

*See Report of 1896–97, Water Commissioners, Clinton, Mass.
†See Report of Bureau of Water, City of Philadelphia, 1896.

11

In the case of water powers, this meter is valuable in determining the quantity of water drawn by tenants of water-rights for power, or for wash water and other purposes other than power.

Special advantages for mills and factories. It offers to mills and factories a means of checking charges for power, or for ascertaining the amount of power used.‡ Figure 5. It can be submerged in a flume or penstock, and enables large bodies of water to be measured regularly and accurately.

4-INCH VENTURI TUBE, SPIGOT ENDS.

MEMORANDA

Column of water 1 foot high ÷ 0.433 lbs. at 62° F.
Column of water 1 foot high Column of Mercury 0.883 ins. high, at 62° F.

1 Gallon	231 cubic ins.
1 "	0.1337 cubic foot.
1 "	8.335 lbs. at 62° F.
1 "	3.786 litres.
1 Cubic foot of water	7.480 gallons.
1 Cubic foot of water	62.355 lbs. at 62° F.

Flow at rate of 1 cubic ft. per second for 24 hours = 646,000 gallons.

$$^2g = 64.33$$
$$\sqrt{^2g} = 8.02$$

‡See Engineering News, Vol. XXXVIII, No. 2, July 8, 1897. "The Plant of the Pioneer Electric Power Co., at Ogden, Utah."

FIGURE 5.

ONE OF TWO 54-INCH VENTURI METERS.
POWER STATION PIONEER ELECTRIC POWER CO.
OGDEN, UTAH.

FIGURE 6.

16-INCH VENTURI METER TUBE.

FIGURE 7.
20-INCH VENTURI METER TUBE.

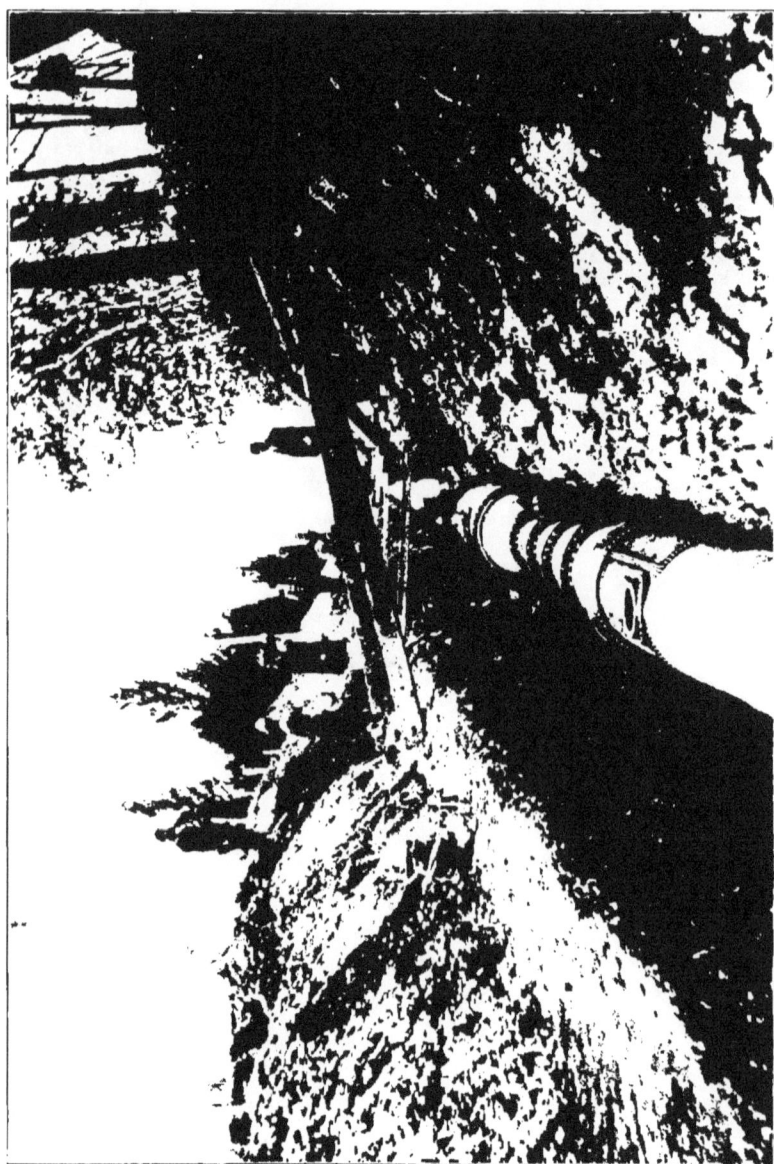

FIGURE 8.

42-Inch Venturi Meter Tube, in Riveted Steel Pipe Line.

FIGURE 9.

48-INCH VENTURI TUBE, IN RIVETED STEEL PIPE LINE.

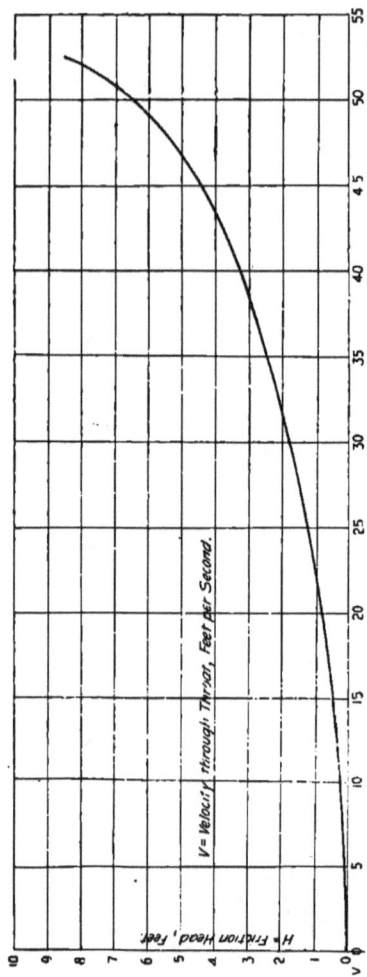

FIGURE 10.

DIAGRAM SHOWING APPROXIMATE LOSS OF HEAD FOR DIFFERENT VELOCITIES OF
FLOW THROUGH THROAT OF VENTURI TUBES.

TABLE SHOWING QUANTITY OF WATER PASSING THROUGH VENTURI METER TUBES OF DIFFERENT SIZES (THROAT AREA 1-9 OF MAIN), WITH CORRESPONDING VELOCITY OF FLOW IN THROAT, " HEAD ON VENTURI," AND " FRICTION HEAD."*

" HEAD ON VENTURI " is the difference of pressure, in feet of water, at throat and up-stream end of tube.

" FRICTION HEAD " is the difference of pressure, in feet of water, at up-stream and down-stream ends of tube, or the LOSS OF HEAD due to introduction of meter tube.

Vel. through throat in ft. per second	Quantity in Cubic Feet per Second.						Head on Venturi, in feet.	Friction Head in feet, approximate.
	10-inch Meter.	12-inch Meter.	15-inch Meter.	16-inch Meter.	18-inch Meter.	20-inch Meter.		
2.5	.152	.218	.340	.389	.490	.608	.097	.015
3.	.182	.261	.408	.466	.589	.728	.14	.02
3.5	.212	.305	.477	.543	.687	.848	.19	.025
4.	.242	.348	.545	.619	.784	.968	.25	.03
5.	.303	.436	.681	.778	.981	1.212	.39	.05
6.	.364	.523	.818	.932	1.179	1.456	.56	.07
7.	.424	.610	.954	1.086	1.374	1.696	.76	.10
8.	.485	.697	1.090	1.238	1.570	1.940	1.00	.13
9.	.545	.785	1.227	1.398	1.767	2.180	1.20	.17
10.	.606	.872	1.362	1.556	1.963	2.424	1.50	.22
12.	.727	1.047	1.636	1.864	2.357	2.908	2.26	.32
14.	.850	1.224	1.908	2.172	2.748	3.400	3.10	.42
16.	.970	1.396	2.181	2.476	3.141	3.880	4.05	.53
18.	1.090	1.570	2.454	2.796	3.534	4.360	5.16	.67
20.	1.212	1.745	2.727	3.112	3.927	4.848	6.40	.82
24.	1.450	2.094	3.272	3.728	4.712	5.800	9.21	1.20
28.	1.700	2.443	3.817	4.344	5.497	6.800	12.73	1.68
32.	1.940	2.792	4.363	4.952	6.287	7.760	17.25	2.10
36.	2.180	3.141	4.908	5.592	7.062	8.720	21.75	2.70
38.	2.300	3.316	5.181	5.913	7.455	9.200	24.50	3.00

* To meet special requirements as to Capacity or Friction Head, Meter Tubes are made with throats of any area less than one-fourth the area of main pipe.

TABLE SHOWING QUANTITY OF WATER PASSING THROUGH VENTURI METER TUBES OF DIFFERENT SIZES (THROAT AREA 1-9 OF MAIN), WITH CORRESPONDING VELOCITY OF FLOW IN THROAT, " HEAD ON VENTURI," AND " FRICTION HEAD."*

" HEAD ON VENTURI" is the difference of pressure, in feet of water, at throat and up-stream end of tube.

" FRICTION HEAD" is the difference of pressure, in feet of water, at up-stream and down-stream ends of tube, or the LOSS OF HEAD due to introduction of meter tube.

Vel. thro' throat in feet per sec.	Quantity in Cubic Feet per second.						Head on Venturi, in feet.	Friction Head in feet, approximate.
	21-inch Meter.	24-inch Meter.	27-inch Meter.	30-inch Meter.	36-inch Meter.	42-inch Meter.		
2.5	.668	.872	1.104	1.363	1.963	2.672	.097	.015
3.	.801	1.046	1.325	1.636	2.355	3.207	.14	.02
3.5	.935	1.221	1.546	1.908	2.748	3.741	.19	.025
4.	1.069	1.396	1.767	2.181	3.141	4.277	.25	.03
5.	1.336	1.744	2.208	2.727	3.927	5.345	.39	.05
6.	1.603	2.092	2.650	3.272	4.712	6.414	.56	.07
7.	1.871	2.443	3.092	3.817	5.497	7.482	.76	.10
8.	2.138	2.792	3.534	4.363	5.283	8.552	1.00	.13
9.	2.405	3.138	3.976	4.908	7.068	9.621	1.20	.17
10.	2.672	3.488	4.417	5.454	7.854	10.690	1.50	.22
12.	3.204	4.188	5.301	6.545	9.424	12.828	2.26	.32
14.	3.740	4.888	6.184	7.630	10.995	14.964	3.10	.42
16.	4.276	5.585	7.068	8.727	12.566	17.104	4.05	.53
18.	4.812	6.284	7.952	9.817	14.137	19.242	5.16	.67
20.	5.345	6.976	8.835	10.908	15.708	21.386	6.40	.82
24.	6.408	8.377	10.602	13.090	18.849	25.656	9.21	1.20
28.	7.484	9.773	12.370	15.271	21.991	29.932	12.73	1.68
32.	8.552	11.170	14.137	17.453	25.132	34.208	17.25	2.10
36.	9.624	12.566	15.904	19.635	28.274	38.484	21.75	2.70
38.	10.155	13.264	16.776	20.725	29.845	40.622	24.50	3.00

* To meet special requirements as to Capacity or Friction Head, Meter Tubes are made with throats of any area less than one-fourth the area of main pipe.

TABLE SHOWING QUANTITY OF WATER PASSING THROUGH
VENTURI METER TUBES OF DIFFERENT SIZES
(THROAT AREA 1-9 OF MAIN), WITH CORRESPOND-
ING VELOCITY OF FLOW IN THROAT, " HEAD ON
VENTURI," AND " FRICTION HEAD."*

"HEAD ON VENTURI" is the difference of pressure,
in feet of water, at throat and up-stream end of tube.

" FRICTION HEAD " is the difference of pressure, in feet
of water, at up-stream and down-stream ends of tube, or
the LOSS OF HEAD due to introduction of meter tube.

Velocity through throat in feet per second.	Quantity in Cubic Feet per Second.					Head on Venturi, in feet.	Friction Head in feet, approximate.
	48-inch Meter.	54-inch Meter.	60-inch Meter.	72-inch Meter.	80-inch Meter.		
2.5	3.490	4.417	5.454	7.854	9.647	.097	.015
3.	4.188	5.301	6.545	9.435	11.577	.14	.02
3.5	4.886	6.185	7.640	10.995	13.507	.19	.025
4.	5.585	7.068	8.728	12.682	15.500	.25	.03
5.	6.981	8.835	10.906	15.708	19.295	.39	.05
6.	8.377	10.602	13.090	18.849	23.154	.56	.07
7.	9.772	12.370	15.280	21.990	27.014	.76	.10
8.	11.170	14.136	17.452	23.364	31.000	1.00	.13
9.	12.564	15.903	19.635	25.305	34.731	1.20	.17
10.	13.962	17.670	21.816	31.416	38.390	1.50	.22
12.	16.754	21.204	26.180	37.698	46.308	2.26	.32
14.	19.554	24.740	30.560	43.980	54.028	3.10	.42
16.	22.340	28.272	34.904	50.728	62.000	4.05	.53
18.	25.128	31.806	39.270	56.610	69.462	5.16	.67
20.	27.924	35.340	43.632	62.832	76.780	6.40	.82
24.	33.508	42.408	52.360	75.396	92.616	9.21	1.20
28.	39.088	49.480	61.120	87.960	108.046	12.73	1.68
32.	44.780	56.544	69.808	101.456	124.000	17.25	2.10
36.	50.256	63.612	78.549	113.738	138.924	21.75	2.70
38.	53.052	67.146	82.900	119.442	147.154	24.50	3.00

* To meet special requirements as to Capacity or Friction Head, Meter Tubes are made
with throats of any area less than one-fourth the area of main pipe.

21

DIAGRAM

SHOWING ARRANGEMENT OF THE

THIRTEEN VENTURI METERS

OF THE EAST JERSEY WATER CO.

IN USE SEPTEMBER 1897

To Bayonne

30 meter

To Kearney and Harrison

16 meter

42 meter

Passaic River at Belleville

42 meter

To Jersey City

To Low Service Newark

To Newark

30 meter

To High Service Newark

48 meter

42 meter

16 meter

12 meter

To Bloomfield

To Montclair

20 meter

To Montclair

16 miles long

42 meter acting in either direction

48" Conduit No 1 21 miles long

42" Conduit No 2

To Little Falls Pumping Station

To Paterson

48" Conduit No 2 6 miles long

48 meter

Gate House

Dam

Intake Reservoir

ACCURACY

The accuracy of the meter has been fully demonstrated by numerous tests, and when these have been made with the care that should be exercised in any hydraulic experiment, most satisfactory results have been obtained.

No better demonstration of the accuracy of the Venturi meter can be presented than the continuous performance of thirteen meters on the works of the East Jersey Water Company. That Company has a contract with the City of Newark, N. J., to supply it with not more than 27½ million gallons of water per day. The Company controls the water shed and plant supplying this water, and is allowed to dispose of the balance that the works supply to other cities and towns. In this way it supplies at the present time Jersey City, the City of Bayonne, the Township of Franklin, the Town of Montclair, N. J., and other consumers. All the water is sold by measure, through ten Venturi meters, and daily records are kept of the quantities delivered to the principal consumers, with weekly and monthly records for the smaller consumers. Daily records are also kept of the quantities delivered to the conduits through receiving meters at the intake.

The arrangement of the meters is shown by diagram, Fig. 11, and the following table compiled from official records of the Company shows comparison of Receiving and Selling meters for seventeen months.

From this table it will be seen that in seventeen

THE EAST JERSEY WATER COMPANY,

CLEMENS HERSCHEL, Superintendent.   2 WALL STREET.   NEW YORK, August 25th, 1897.

QUANTITIES STATED IN GALLONS.

DATE.	RECEIVING METERS.	SELLING METERS.	DIFFERENCE. Leakage of pipe, discrepancy of meters, or both.	REMARKS.
1896.				
January .	1,060,200,000	1,064,300,000	— 4,100,000	Conduit No. 1.
February .	1,030,700,000	1,037,600,000	— 6,900,000	21 miles 48 inch pipe.
March .	1,041,100,000	1,040,800,000	+ 300,000	21 miles 48 inch pipe.
April . .	1,103,000,000	1,101,400,000	+ 1,600,000	{ March 30th and 31st, unmeasured draft from pipe line.
May	1,142,600,000	1,138,100,000	+ 4,500,000	21 miles 48 inch pipe.
June .	1,363,000,000	1,337,700,000	+ 25,300,000	21 miles 48 inch pipe.
July .	1,460,000,000	1,445,500,000	+ 14,500,000	{ Large unmeasured draft from Conduit, to test Conduit No. 2.
August	1,486,400,000	1,443,000,000	+ 43,400,000	Same as in June.
September	1,348,000,000	1,329,700,000	+ 18,300,000	Same as in June.
October .	745,800,000 699,200,000 1,445,000,000	1,429,900,000	+ 15,100,000	{ Conduit No. 1, 21 miles 48 inch pipe. Conduit No. 2, 6 miles 48 in. pipe, 16 miles 42 in. pipe, large unmeasured draft to test conduit No. 2.
November	888,000,000 570,600,000 1,458,600,000	1,461,200,000	— 2,600,000	Conduit No. 1. Conduit No. 2.

Month	Conduit No. 1	Conduit No. 2				Remarks
December	941,600,000	760,300,000	1,700,100,000	+ 1,800,000		Conduit No. 1. / Conduit No. 2.
		1,701,900,000				
1897. January	934,600,000	485,600,000	1,777,900,000	+ 2,300,000		Conduit No. 1. / Conduit No. 2.
		1,780,200,000				
February	844,700,000	746,200,000	1,590,400,000	+ 500,000		Conduit No. 1. / Conduit No. 2.
		1,590,900,000				
March	940,700,000	708,600,000	1,646,800,000	+ 2,500,000		Conduit No. 1. / Conduit No. 2.
		1,649,300,000				
April	997,100,000	650,300,000	1,552,000,000	+ 5,400,000		Conduit No. 1. / Conduit No. 2.
		1,557,400,000				
May	942,700,000	684,900,000	1,622,000,000	+ 5,600,000		Conduit No. 1. / Conduit No. 2.
		1,627,600,000				
June	889,100,000	741,500,000	1,628,500,000	+ 2,100,000		Conduit No. 1. / Conduit No. 2.
		1,630,600,000				
July	846,100,000	896,100,000	1,734,800,000	+ 7,400,000		Conduit No. 1. / Conduit No. 2. Large unmeasured draft from the Conduits, to test Paterson Extension.
		1,742,200,000				

months 27,218,700,000 gallons of water were delivered into the conduits Nos. 1 and 2, through two 48-inch intake meters, and remeasured through ten selling or outlet meters, varying in size from 12 to 48 inches, with a difference of measurements between the two sets of meters of only ½ of 1 per cent. Considering only the months November, 1896, to July, 1897, during which performance of the meters was not interfered with by irregular " unmeasured drafts of water " for testing pipe lines, etc., it will be seen that 12,996,500,000 gallons of water were measured by the intake meters and remeasured by the selling meters, with a difference of only 17,600,000 gallons, or 14-100 of 1 per cent.

12-INCH VENTURI METER TUBE, SPIGOT ENDS.

SETTING OF METER

The meter tube is set in the pipe-line, wherever most convenient. See Figures 5, 6, 7, 8 and 9. The register is usually placed ten feet or more below the hydraulic grade, and not more than 1000 feet from the tube. The tube and register are connected by two lines of ½ inch brass, lead or tin-lined pipe, and as a matter of economy are usually placed as near one another as possible.

The register must be properly protected from freezing, and when a gate-house, pumping-station or other building suitable for the purpose is not available a vault or register house must be provided. This should be frost proof, and not less than 6 ft. x 6 ft. inside; but in other respects may be built to suit the taste and requirements of the purchaser. Figures 12, 13 14, 15 and 16 illustrate a few that have been found entirely satisfactory. Drawings for that shown by Figure 14 will be furnished when desired.

When the meter must be placed where frequent readings cannot easily be obtained, the registrations may be automatically transmitted by electricity to a secondary or office dial, figure 17, which may be placed any distance from the register.

FIGURE 12.

REGISTER HOUSE, WORCESTER, MASS.

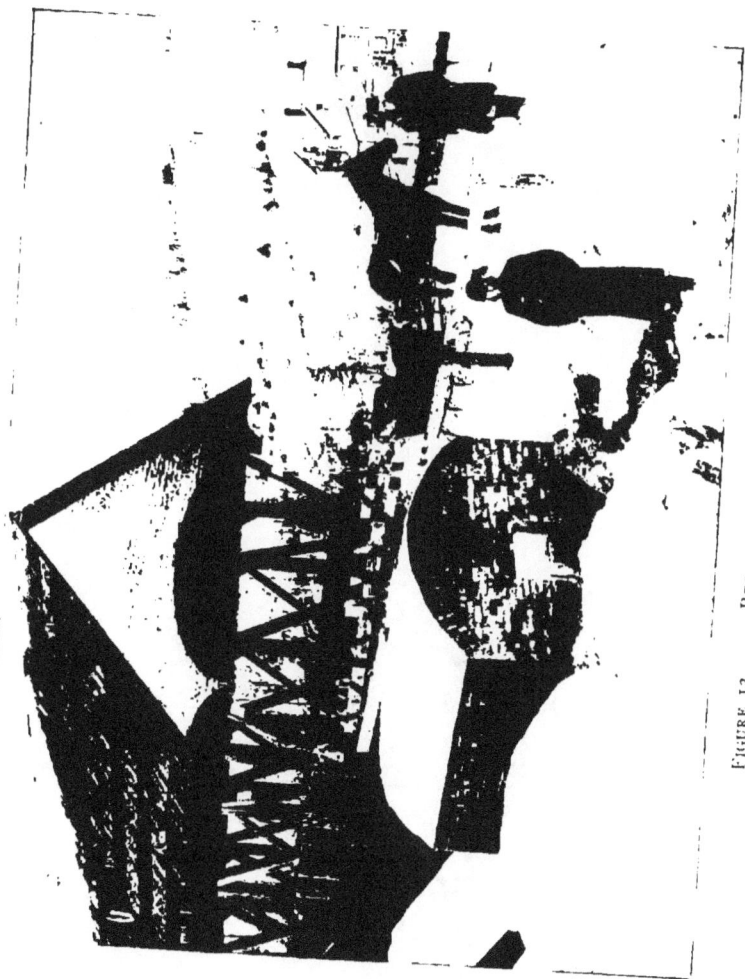

FIGURE 13.   REGISTER HOUSE, ST. JOHNSBURY, VT.

FIGURE 14.   REGISTER HOUSE, READING, PA.

FIGURE 15. REGISTER HOUSE, ST. PAUL, MINN.

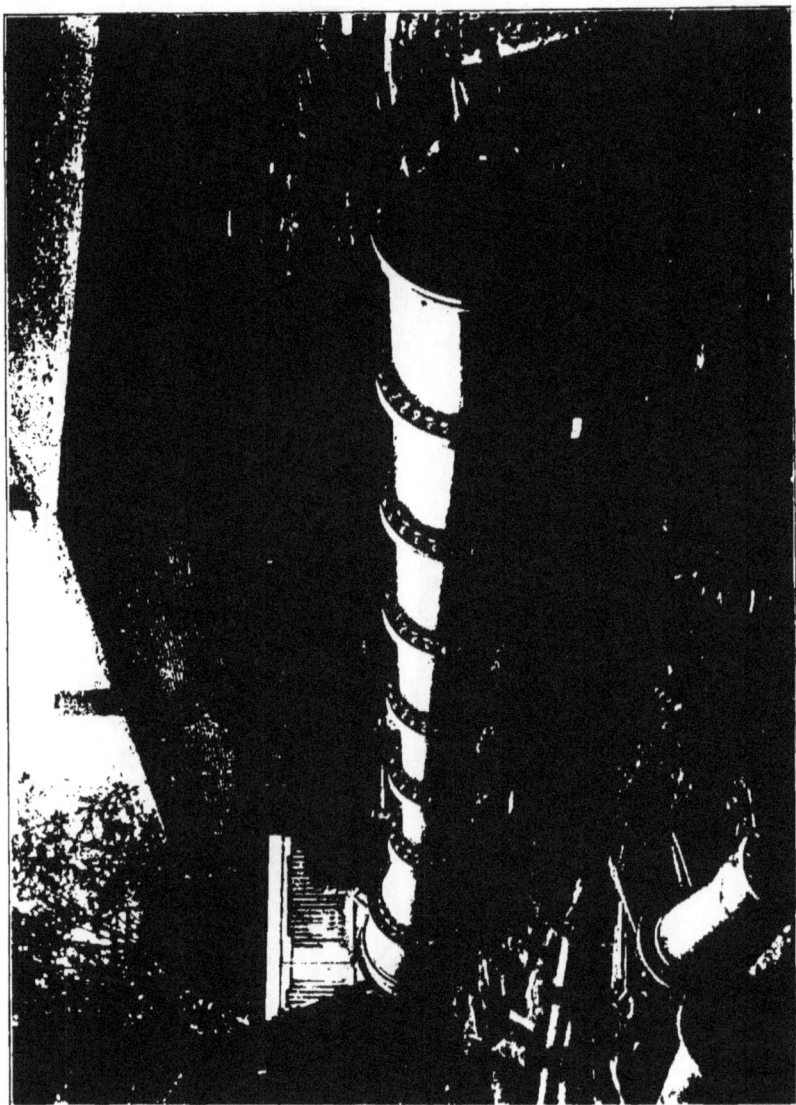

60-Inch Venturi Meter Tube. Flange Ends.

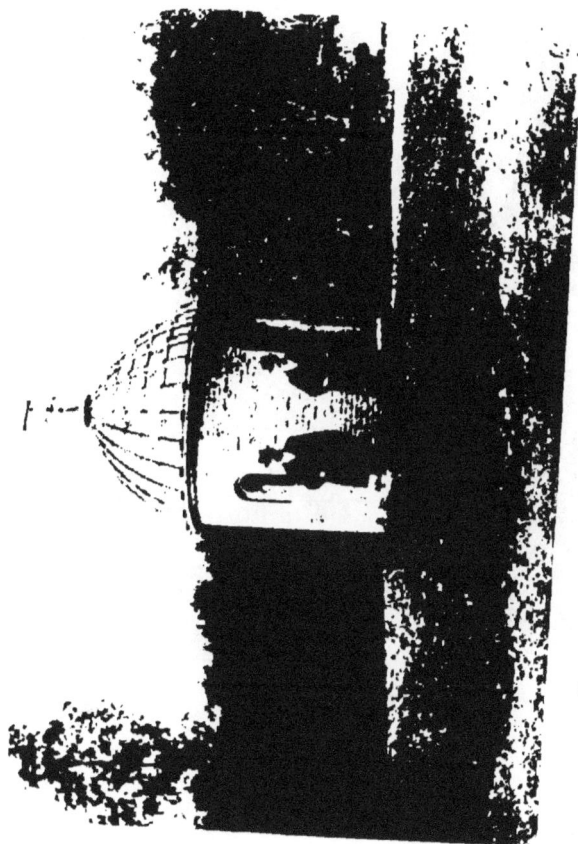

FIGURE 16.   REGISTER HOUSE, WORCESTER, MASS.

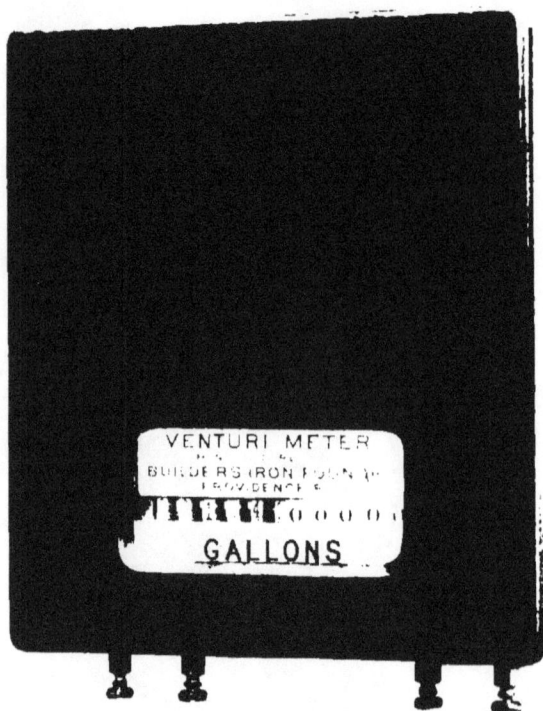

VENTURI METER

BUILDERS IRON FOUNDRY
PROVIDENCE R

GALLONS

FIGURE 17.

SECONDARY OR OFFICE DIAL.

THE VENTURI WATER METER.

MANSFIELD MERRIMAN'S "HYDRAULICS."

ARTICLE 71.

" It has been shown by Herschel* that a compound tube provided with piezometers may be used for the accurate measurement of water. The apparatus, which is called by him the Venturi Water Meter, is shown in outline in the accompanying figure, and consists of a compound tube terminated by cylinders, into the top of which are tapped

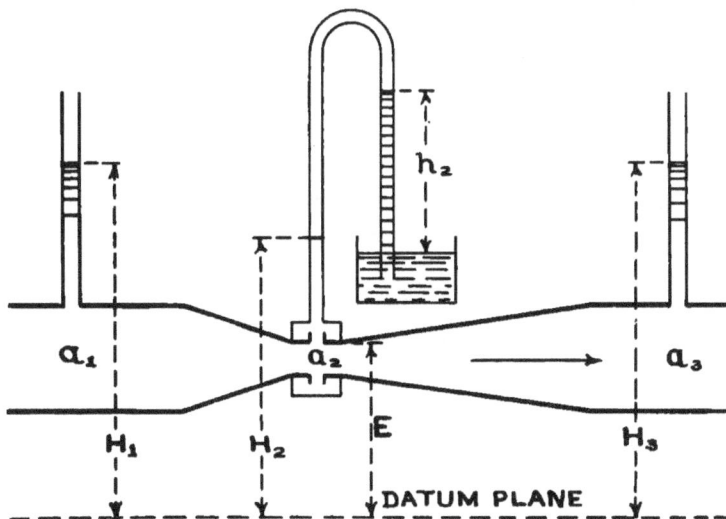

the piezometers H_1 and H_3. Surrounding the small section a_2 is a chamber into which four or more holes lead from the top, bottom and sides of the tube, and from

---

* Transactions American Society of Civil Engineers, 1887, Vol. XVIII, p. 228.

which rises the piezometer H_2. The flow passing through the tube has the velocities v_1, v_2, and v_3 at the sections a_1, a_2, and a_3, and these velocities are inversely as the areas of the sections. When the pressure in a_2 is positive, the water stands in the central piezometer at a height H_2, as shown in the figure; when the pressure is negative the air is rarefied, and a column of water lifted to the height h_2. If E is the height of the top of the section a_2 above the datum, the value of H_2 for the case of negative pressure was taken to be $E-h_2$. The apparatus was constructed so that the areas a_1 and a_3 were equal, while a_2 was about 1-9 of these.

To determine the discharge per second through the tube, the areas a_1 and a_2 are to be accurately found by measurements of the diameters"; then (the quantity passing is equal to the area X the velocity or)

$$Q = a_1\, v_1, \text{ or } Q = a_2\, v_2.$$

If no losses of head due to friction occur between the sections a_1 and a_2, the quantity h′ in the formula of the last article is 0, and

$$0 = \frac{v_1^{\,2} - v_2^{\,2}}{2g} + H_1 - H_2. \,\dagger$$

Inserting in this for v_1 and v_2 their values in terms of Q, and then solving for Q, gives the result

$$Q = \frac{a_1\, a_2}{\sqrt{a_1^{\,2} - a_2^{\,2}}} \sqrt{2g(H_1 - H_2)}.$$

which may be called the theoretic discharge. Dividing this expression by a_1 gives the velocity v_1, and dividing it

† This equation is deduced from the well-known law that the sum of velocity and friction heads is constant.

by a_2 gives the velocity v_2. Owing to the losses of head which actually exist, this expression is to be multiplied by a co-efficient c; thus:

$$q = c. \frac{a_1\, a_2}{\sqrt{a_1^2 - a_2^2}} \sqrt{2g\,(H_1 - H_2\,)}$$

is the formula for the actual discharge per second.

Reference is made to Herschel's paper, above quoted, for a full description of the method of conducting the experiments. The discharge was actually measured either in a large tank or by a weir; and thus q being known for observed piezometer heights H_1 and H_2, the value of c was computed by dividing the actual by the theoretic discharge. For example, the smaller tube used had the areas

$$a_1 = 0.77288, \quad a_2 = 0.08634 \text{ square feet};$$

hence the theoretic discharge is

$$Q = 0.086884 \ \sqrt{2g\,(H_1 - H_2\,)},$$

and the co-efficient of discharge or velocity is

$$c = \frac{q}{Q}.$$

In experiment No. 1 the value of H_1 was 99.069, while h_2 was 24.509 feet, and the actual discharge was 4.29 cubic feet per second. As E was 84.704, the value of H_2 is 60.195 feet. The theoretic discharge then is

$$Q = 0.086884 \times 8.02 \sqrt{38.874} = 4.345.$$

Dividing 4.29 by this, gives for c the value 0.988. Fifty-five experiments made in this manner, in all of which negative pressure existed in a_2, gave co-efficients ranging in value from 0.94 to 1.04, only four being greater than 1.01 and only two less than 0.96.

The larger tube used had the areas $a_1 = 57.823$ and $a_2 = 7.074$ square feet, and the pressure at the central piezometer was both positive and negative. Twenty-eight experiments give co-efficients ranging from 0.95 to 0.99, the highest co-efficients being for the lowest velocities. In this tube the velocity at the section a_2 ranged from 5 to 34.5 feet per second. The small variation in the co-efficients for the large range in velocity indicates that the apparatus may in the future take a high rank as an accurate instrument for the measurement of water. Under low velocities, however, it is not probable that the arrangement of piezometers shown in the accompanying figure will give the best results; in order that H_1 may correctly indicate the mean pressure in a_1, connection seems to be required both at the bottom and sides of the tube like that at a_2. It is thought, moreover, that the elevation E should be measured to the centre of the section rather than to the top. The lower piezometer H_3 is not an essential part of the apparatus and may be omitted, although it was of value in the experiments as showing the total loss of head.

www.ingramcontent.com/pod-product-compliance
Lightning Source LLC
Chambersburg PA
CBHW021451090426
42739CB00009B/1722